10-Minute Critical-Thinking Activities for Math

by
Hope Martin

J. WESTON
WALCH
PUBLISHER

Portland, Maine

User's Guide to *Walch Reproducible Books*

As part of our general effort to provide educational materials that are as practical and economical as possible, we have designated this publication a "reproducible book." The designation means that the purchase of the book includes purchase of the right to limited reproduction of all pages on which this symbol appears:

Here is the basic Walch policy: We grant to individual purchasers of this book the right to make sufficient copies of reproducible pages for use by all students of a single teacher. This permission is limited to a single teacher and does not apply to entire schools or school systems, so institutions purchasing the book should pass the permission on to a single teacher. Copying of the book or its parts for resale is prohibited.

Any questions regarding this policy or request to purchase further reproduction rights should be addressed to:

Permissions Editor
J. Weston Walch, Publisher
321 Valley Street • P.O. Box 658
Portland, Maine 04104-0658

1 2 3 4 5 6 7 8 9 10
ISBN 0-8251-3816-7
Copyright © 1998
J. Weston Walch, Publisher
P. O. Box 658 • Portland, Maine 04104-0658
Printed in the United States of America

Contents

To the Teacher .... *iv*

The Activities: Matrix of Critical-Thinking Skills .... *vi*

PART 1: Critical Thinking and Algebra 1

PART 2: Logic and Critical Thinking 13

PART 3: Number Theory and
Problem Solving 25

PART 4: Sequences, Patterns, and Codes 37

PART 5: Visual/Geometric Patterns 51

Answers *65*

To the Teacher

What does it mean to *think critically* in the mathematics classroom? Critical thinking is disciplined, self-directed thinking where students are asked to employ previously learned skills and concepts in a problem-solving setting. The problems and activities in *10-Minute Critical-Thinking Activities for Math* have been designed to (1) develop students' reasoning and thinking skills, (2) encourage them to analyze their results, and (3) require students to justify or reexamine their procedures.

However, problems, in and of themselves, cannot be defined as critical-thinking problems. It may be the type of activity that encourages students to reason, analyze, make modifications, and reexamine their procedures. Then we, as teachers, must tempt our students to become actively involved in their learning. These warm-ups support active learning by asking students to explain their reasoning and share their strategies with the class, encouraging them to develop orginal problems and share them with the class, and inviting them to reevaluate their work and check for accuracy.

The activities in *10-Minute Critical-Thinking Activities for Math* contain non-traditional problems designed to encourage students to use critical or logical thinking skills. They have been grouped into five sections:

1. Critical Thinking and Algebra;

2. Logic and Critical Thinking;

3. Number Theory and Problem Solving;

4. Sequences, Patterns, and Codes; and

5. Visual/Geometric Patterns.

In each section the activities are arranged progressively from the easiest to the most difficult. However, do not feel you must complete all of the activities in a section before moving on. Any problem can be used at any time; no sequence of skills is assumed. By selecting problems from different sections, teachers can bring variety into the classroom and revisit topics throughout the year.

Many of the problems in this book are considered open-ended—they may have one correct answer but more than one way to find it, or the problem may actually have more than one correct answer. Open-ended problems encourage students to try different methods of solution and support divergent strategies and a variety of learning styles. When students are encouraged to find multiple solutions using a variety of strategies, not only do they become better problem solvers, but mathematics becomes a creative expression of their talents.

The activities in *10-Minute Critical-Thinking Activities for Math* provide mathematical enrichment for students while also addressing topics that are a meaningful part of the mathematics curriculum. These include the following:

- Number theory
- Computation skills
- Geometry concepts and skills
- Mathematical reasoning
- Sequencing and patterning
- Order of operations
- Algebra concepts and skills
- Spatial visualization and transformations

Some teachers may wish to use these masters to make overhead transparencies; other may wish to give students their own copy.

So use the first 10 minutes of your teaching period to expand your students' mathematical horizons (and knowledge) with these stimulating warm-ups! Remember, 10 minutes a day is 1,800 minutes a year, or about 9 weeks of classes each year! Enjoy mathematics each and every minute!

The Activities

Matrix of Critical-Thinking Skills

Page No.	ACTIVITY	Problem Solving	Mathematical Reasoning	Mathematical Communication	Generalization	Visual Thinking	Logic	Patterns & Sequences	Analytic Computation	Analysis	Open-Ended Problems
2	He Said—She Said 1	✔	✔		✔			✔	✔	✔	✔
3	He Said—She Said 2	✔	✔		✔			✔	✔	✔	✔
4	What Comes Next 1	✔	✔		✔	✔		✔	✔	✔	✔
5	What Comes Next 2	✔	✔		✔	✔		✔	✔	✔	✔
6	What Comes Next 3	✔	✔		✔	✔		✔	✔	✔	✔
7	What Comes Next 4	✔			✔	✔		✔	✔	✔	✔
8	The Ladybug Concert	✔	✔	✔			✔		✔	✔	✔
9	The Peanut Problem	✔	✔	✔			✔		✔	✔	✔
10	Number Magic 1	✔	✔	✔			✔		✔		✔
11	Number Magic 2	✔	✔	✔			✔		✔		✔
12	Number Magic 3	✔	✔	✔			✔		✔		✔
14	Ladder Logic	✔	✔	✔	✔	✔	✔				✔
15	Dinner Table Logic	✔	✔	✔		✔	✔				✔
16	Fun with Logic 1	✔	✔	✔		✔	✔				✔
17	Fun with Logic 2	✔	✔	✔		✔	✔				✔
18	Fun with Logic 3	✔	✔	✔		✔	✔				✔
19	Splatland 1	✔	✔	✔	✔		✔		✔	✔	✔
20	Splatland 2	✔	✔	✔	✔		✔		✔	✔	✔
21	Venn Logic 1	✔	✔	✔		✔	✔	✔	✔	✔	✔
22	Venn Logic 2	✔	✔	✔		✔	✔	✔	✔	✔	✔
23	Venn Logic 3	✔	✔	✔		✔	✔	✔	✔	✔	✔
26	Hit the Target 1	✔	✔			✔		✔	✔	✔	✔
27	Hit the Target 2	✔	✔			✔		✔	✔	✔	✔
28	Hit the Target 3	✔	✔			✔		✔	✔	✔	✔
29	The 5 by 5 Array	✔	✔			✔	✔	✔			✔
30	Marbles	✔	✔		✔	✔	✔		✔	✔	✔
31	Triangular Sums	✔	✔		✔	✔		✔	✔	✔	✔

(continued)

Matrix of Critical-Thinking Skills *(continued)*

Page No.	Activity	Problem Solving	Mathematical Reasoning	Mathematical Communication	Generalization	Visual Thinking	Logic	Patterns & Sequences	Analytic Computation	Analysis	Open-Ended Problems
32	A Moving Experience 1	✔	✔		✔	✔	✔	✔		✔	✔
33	A Moving Experience 2	✔	✔		✔	✔	✔	✔		✔	✔
34	Card Tricks 1	✔	✔			✔		✔	✔	✔	✔
35	Card Tricks 2	✔	✔			✔		✔	✔	✔	✔
36	Card Tricks 3	✔	✔			✔		✔	✔	✔	✔
38	Sequences	✔	✔		✔	✔	✔	✔	✔		
39	What Comes Next? 1	✔	✔		✔	✔	✔	✔		✔	
40	What Comes Next? 2	✔	✔		✔	✔	✔	✔		✔	
41	Number Patterns 1	✔	✔		✔			✔	✔	✔	
42	Number Patterns 2	✔	✔		✔			✔	✔	✔	
43	Crack the Code 1	✔	✔		✔	✔	✔	✔	✔	✔	✔
44	Crack the Code 2	✔	✔		✔	✔	✔	✔	✔	✔	✔
45	Crack the Code 3	✔	✔		✔	✔	✔	✔	✔	✔	✔
46	Calendar Math	✔	✔			✔	✔	✔	✔	✔	✔
47	Addition Table Patterns	✔	✔		✔	✔	✔	✔	✔	✔	✔
48	Groups of Numbers 1	✔	✔				✔	✔	✔	✔	
49	Groups of Numbers 2	✔	✔				✔	✔	✔	✔	
53	How Many Triangles?	✔	✔			✔		✔	✔	✔	✔
54	The Checkerboard	✔	✔	✔	✔	✔		✔	✔	✔	✔
55	How Many Triangles in the Pentagon?	✔	✔		✔	✔		✔	✔	✔	✔
56	How Many Lines in This Hexagon?	✔	✔		✔	✔		✔	✔	✔	✔
57	The Puzzling Cube 1	✔	✔			✔		✔		✔	✔
58	The Puzzling Cube 2	✔	✔			✔		✔		✔	✔
59	Paper Folds 1	✔	✔			✔		✔		✔	✔
60	Paper Folds 2	✔	✔			✔		✔		✔	✔
61	Patterns in Geometric Numbers 1	✔	✔		✔	✔	✔	✔	✔	✔	✔
62	Patterns in Geometric Numbers 2	✔	✔		✔	✔	✔	✔	✔	✔	✔
63	Tangram Puzzler 1	✔	✔	✔	✔	✔		✔	✔	✔	✔
64	Tangram Puzzler 2	✔	✔	✔	✔	✔		✔	✔	✔	✔

PART 1: Critical Thinking and Algebra

As students advance in school, they are asked to move from working only with the concrete concepts associated with arithmetic to understanding and using the more abstract concepts of algebra. Students should be given the opportunity to use manipulatives, solve interesting problems with real-world applications, and apply reasoning and problem-solving skills to algebra-type problems.

To start off, "He Said—She Said" problems make functions and the use of variables interesting and fun for students. They should be encouraged to develop their own versions of this game and share them with the class.

"What Comes Next?" activities will help students develop critical-thinking skills and an understanding of the value and power of algebra by observing a variety of interesting visual patterns. These problems help students move from a situation where counting might be an appropriate strategy to a point where forming a generalized rule for the pattern is necessary. By moving from the concrete to a generalization, students develop mathematical thinking and reasoning skills.

Guess and check as a problem-solving strategy is appropriate at times, but it can be very frustrating. The types of problems modeled by "The Ladybug Concert" and "The Peanut Problem" encourage students to experiment with and apply different ways of looking for solutions. At the same time, they are expanding their understanding of algebra. Students should be encouraged to develop similar problems and present them to the class.

The concept of variables is very difficult for many students. The type of problem presented in the several pages of "Number Magic" fascinates students and provokes the question, "How did you know that?" *Is it magic or algebra?*

All of the activities in this chapter encourage students to practice previously learned problem-solving strategies while stretching their minds and asking them to look at alternatives to more comfortable approaches. As their mathematics and reasoning skills grow, students develop self-assurance and become more confident problem solvers.

He Said—She Said 1

Joe **Maria**

Joe and Maria were playing the game, *He Said—She Said.* Joe calls out a number, and Maria tells him what the number turns into. Then it becomes Joe's job to figure out what rule Maria has used to get her answer. For example, Joe says **1** and Maria says **4**; Joe says **3** and Maria says **6**; Joe says **9** and Maria says **12**. Joe knows the answer: He says, "The rule is to add 3." Use the clues below to help you find a new rule.

Joe Says	Maria Says
2	5
6	13
4	9
0	1
What is Maria's rule?	

Now make up your own rule: _____

He Said—She Said 2

Joe

Maria

Joe and Maria were playing the game, *He Said—She Said*. Joe calls out a number, and Maria tells him what the number turns into. Then it becomes Joe's job to figure out what rule Maria has used to get her answer. For example, Joe says **1** and Maria says **4**; Joe says **3** and Maria says **6**; Joe says **9** and Maria says **12**. Joe knows the answer: He says, "The rule is to add 3." Use the clues below to help you find a new rule.

Joe Says	Maria Says
0	−4
6	8
4	4
1	−2
What is Maria's rule?	

Now make up your own rule: _____

© 1998 J. Weston Walch, Publisher   3   *10-Minute Critical-Thinking Activities for Math*

Name _____ Date _____

What Comes Next? 1

Directions: Examine the triangle pattern shown below. See if you can discover a relationship between the first and the second, the first and the third, and so on. How many toothpicks will there be in the fourth pattern? the fifth? the tenth? Can you figure out the answer without counting the toothpicks? Can you find the twentieth in the pattern? The fiftieth? The one-hundredth? Use your algebra skills to help you find the pattern for a sequence with *n* number of terms.

△	◇	◇◇			
1	2	3	4	5	10

1. The twentieth in the sequence will look like this:

   It will contain _____ toothpicks.

2. The fiftieth in the sequence will contain _____ toothpicks.

3. The one-hundredth will contain _____ toothpicks.

4. Write a formula that you could use to calculate the *n*th term

   in the sequence. _____

© 1998 J. Weston Walch, Publisher   4   10-Minute Critical-Thinking Activities for Math

What Comes Next? 2

Directions: Examine the pattern made with squares shown below. See if you can discover a relationship between the first and the second, the first and the third, and so on. How many squares will there be in the fourth pattern? the fifth? the tenth? Can you figure out the answer without counting the squares? Can you find the twentieth in the pattern? The fiftieth? The one-hundredth? Use your algebra skills to help you find the pattern for a sequence with *n* number of terms.

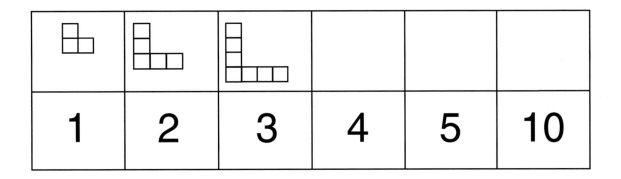

1. The twentieth in the sequence will look like this:

    It will contain _____ squares.

2. The fiftieth in the sequence will contain _____ squares.

3. The one-hundredth will contain _____ squares.

4. Write a formula that you could use to calculate the *n*th term in the sequence. _____

Name _____   Date _____

What Comes Next? 3

Directions: Examine the arrangement of squares shown below. See if you can discover a relationship between the first and the second, the first and the third, and so on. How many squares will there be in the fourth pattern? the fifth? the tenth? Can you figure out the answer without counting the squares? Can you find the twentieth in the pattern? The fiftieth? The one-hundredth? Use your algebra skills to help you find the pattern for a sequence with *n* number of terms.

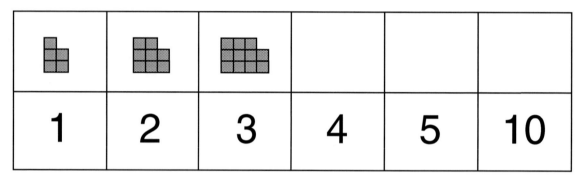

1. The twentieth in the sequence will look like this:

    It will contain _____ squares.

2. The fiftieth in the sequence will contain _____ squares.

    Describe what it will look like. _____

3. The one-hundredth will contain _____ squares.

    Describe what it will look like. _____

4. Write a formula that you could use to calculate the *n*th term

    in the sequence. _____

Name _____   Date _____

What Comes Next? 4

Directions: Examine the arrangement of squares shown below. See if you can discover a relationship between the first and the second, the first and the third, and so on. How many squares will there be in the fourth pattern? the fifth? the tenth? Can you figure out the answer without counting the squares? Can you find the twentieth in the pattern? The fiftieth? The one-hundredth? Use your algebra skills to help you find the pattern for a sequence with *n* number of terms.

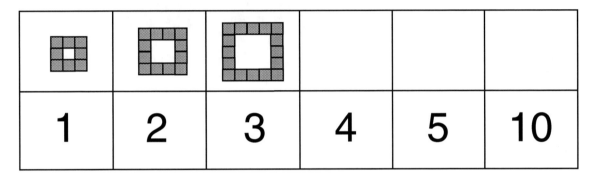

1. The twentieth in the sequence will look like this:

    It will contain _____ squares.

2. The fiftieth in the sequence will contain _____ squares.

    Describe what it will look like. _____

3. The one-hundredth in the sequence will contain _____ squares.

    Describe what it will look like. _____

4. Write a formula that you could use to calculate the *n*th term

    in the sequence. _____

The Ladybug Concert

The Ladybugs are coming to town for a concert. There is such excitement that the concert hall has set up six lines at the box office.

Interestingly, the first line is very, very long, but the second line has one more person than the first line, and the third line has one more person than the second line, and so on. Each of the lines has one more person than the one before it. Altogether, there are 327 people in all six lines. How many people are in each of the six lines?

Explain how you solved the problem.

The Peanut Problem

In a package of mixed nuts, Marty found the following:

- There were twice as many Brazil nuts as cashews;
- There were twice as many peanuts as Brazil nuts;
- There were the same number of almonds as Brazil nuts.

Altogether, there were 81 nuts in the tin. How many of each type of nut were there?

Brazil nuts: _____

Cashews: _____

Almonds: _____

Peanuts: _____

Number Magic 1

We know what's on your mind! Follow these directions, and see if you can figure out our magic trick!

- ⇨ Think of a number—write it down.
- ⇨ Add 2.
- ⇨ Multiply by 4.
- ⇨ Divide it in half
- ⇨ Subtract 4.
- ⇨ Divide by the number you started with.
- ⇨ Your result is 2.

(*Hint:* Use your algebra skills to unravel the magic.)

Name _____ Date _____

Number Magic 2

We know what's on your mind! Follow these directions, and see if you can figure out our magic trick!

⇨ Pick a number between 1 and 10—write it down.

⇨ Multiply by 6.

⇨ Add 4.

⇨ Divide by 2.

⇨ Subtract 2.

⇨ Divide by 3.

⇨ You should have the number you started with.

(*Hint:* Use your algebra skills to unravel the magic.)

Number Magic 3

We know what's on your mind! Follow these directions, and see if you can figure out our magic trick!

⇨ Pick a number—write it down.

⇨ Add 5.

⇨ Multiply by 5.

⇨ Subtract 25.

⇨ Divide by the number you started with.

⇨ You are left with the number 5.

(*Hint:* Use your algebra skills to unravel the magic.)

PART 2: Logic and Critical Thinking

The development of logical reasoning is fundamental to learning mathematics. Logic problems stretch the brain and help students develop thinking and problem-solving skills. The NCTM standards state that students should have the opportunity to recognize and apply deductive and inductive reasoning. On page 81 the standards document recommends that students "need a great deal of time and many experiences to develop their ability to construct valid arguments in problem settings and evaluate the arguments of others."

The first set of problems—"Ladder Logic"—contains no numbers, so . . . how can they be mathematics? These problems require students to think visually and to plan ahead, both important problem-solving strategies in mathematics. When students have solved these, encourage them to develop their own ladders and share them with the class. By working backwards, students continue to strengthen and develop their ability to solve problems.

"Fun with Logic" can be used to develop students' mathematical reading skills. Reading mathematics requires different skills and strategies from reading prose. Mathematics reading material is denser: Statements often contain many more individual units or ideas than the prose students are used to, and more information per square inch requires a slower reading pace and more attention to detail. Also, because of the complexity of logic problems, breaking them down into smaller problems is sometimes necessary—a strategy that can then be used to solve other types of problems. Matrix logic problems give students experience reading mathematical clues and applying them to known information. They become a fun way for students to experience deductive reasoning.

The two "Splatland" problems offer interesting challenges to develop both algebra and logical reasoning skills. Again, students must read mathematically and use a step-by-step problem-solving approach.

Venn diagrams require special attention in the classroom. Such terms as *at least, and, either, or,* and *only* can be very confusing to students, and it cannot be assumed that they will understand the meaning of these terms within the context of a logic problem. In "Venn Logic 1," students need to know the attributes of the geometric shapes to decide which section of the Venn diagram best describes each one. "Venn Logic 2" adds a new twist: Does the statement, "19 students play baseball," mean that they play *only* baseball? As students read through the problem, they see that the answer to this question is "No." Nine students play both baseball and soccer, and six play baseball, soccer, and tennis; they should therefore be placed in the center of the three circles. Students need to understand what each section of each circle stands for. "Venn Logic 3" focuses on the significance of *and, only, or,* and so forth. These little words are often overlooked in the reading of prose but are important in understanding the Venn diagram.

Ladder Logic

You can change one letter on each step, and each step has to contain a **real word**. You can move up and down the ladder.

point

raids

---

bolt

card

See if you can design your own ladder logic problems!

Dinner Table Logic

Juan is planning a dinner party for himself and seven of his friends. Read these clues carefully and place Juan and his friends in their correct places around the dinner table.

1. Scott is seated between Eliza and Marta and across from Joey; Marta is seated to Scott's left.

2. Eliza is seated to the right of Scott, to the left of Juan and directly across from Christine.

3. Juan is seated directly across from Charlie and next to Jennifer.

4. Juan has arranged for boys and girls to alternate around the table, i.e., boy, girl, boy, girl, etc.

5. Every girl is seated directly across from every other girl; Marta is seated directly across from Jennifer.

6. Jennifer is to the right of Juan and to the left of Joey.

7. Christine is seated to the right of Joey and to the left of Charlie.

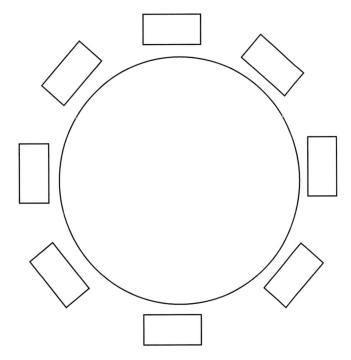

Name _____ Date _____

Fun with Logic 1

Amy, Burt, Charlie, Donna, and Eliza all went to school together. Their professions: architect, botanist, chemist, doctor, and electrician. Figure out the profession of each. You can use the matrix to help you sort out the clues.

1. None of the classmates works in a job that begins with the same letter as their name.
2. Burt and Donna (who are married to each other) called the electrician to rewire their apartment.
3. The chemist went to cheer on the doctor and Burt, who play on the same softball team.
4. Eliza has an allergy to pollen, and Charlie faints at the sight of blood.
5. The unmarried electrician is dating Charlie.

	Architect	Botanist	Chemist	Doctor	Electrician
Amy					
Burt					
Charlie					
Donna					
Eliza					

© 1998 J. Weston Walch, Publisher   16   10-Minute Critical-Thinking Activities for Math

Fun with Logic 2

1. Anthony, Brandy, Milly, Sam, and Polly attend the same school. They each have a favorite subject but it does not begin with the same first letter as their name.

2. Brandy and the person whose favorite subject is Social Studies have lunch together every day.

3. Polly and Sam both dislike the experiments in science class.

4. Molly is a very talented athlete and is on the track and volleyball teams.

5. Sam jokingly says, "I can't draw a straight line with a **ruler!**"

Can you problem-solve which student liked which subject?

	Art	Biology	Math	Social Studies	Physical Education
Anthony					
Brandy					
Molly					
Sam					
Polly					

Name _____ Date _____

Fun with Logic 3

1. Pat, Charlie, Luan, Maria, and Josie all take piano lessons from the same teacher, Mrs. Wang, but none of them go on the same day.
2. Maria and Josie play tennis together on Tuesday and Maria has soccer practice on Monday.
3. Charlie takes his lessons on Tuesday.
4. Luan and Pat met Maria on Thursday; she was on her way to her lesson.
5. Pat and Charlie had play rehearsal on Wednesday; one of them missed her or his lesson.
6. Luan tells Charlie what he will need to practice for his lesson.

Who took piano lessons on which day?

	Monday	Tuesday	Wednesday	Thursday	Friday
Pat					
Charlie					
Luan					
Maria					
Josie					

Splatland 1

A Dab **A Dud** **A Dib** **A Doozie**

In Splatland there are four different creatures, Dibs, Dabs, Duds, and Doozies. Each of the residents of Splatland has a different weight, but the mayor has found numerical relationships among them. These are clues to the weights:

- Two Dibs weigh the same as one Dab.
- One and one-half Dabs weigh the same as one Doozie.
- One Dud weighs the same as one-half a Dib.
- One Doozie weighs 12 grams.

Find the weight of each Dib, Dab, and Dud.

Dib: _____

Dab: _____

Dud: _____

Splatland 2

    A Dab   A Dud   A Dib   A Doozie

In Splatland there are four different creatures, Dibs, Dabs, Duds, and Doozies. Each of the residents of Splatland has a different weight, but the mayor has found numerical relationships among them. These are clues to the weights:

- One Doozie weighs twice as much as a Dib.
- One Dud is one-third the weight of one Dib.
- One Dab weighs 4 grams.
- Altogether they weigh 34 grams.

Find the weight of each Dib, Doozie, and Dud.

Dib: _____

Doozie: _____

Dud: _____

Venn Logic 1

Below is a list of geometric shapes. Place them in the section of the Venn Diagram that best describes the shape.

square   trapezoid   rhombus
equilateral triangle   isosceles triangle   isosceles right triangle
rectangle   parallelogram   right trapezoid
right triangle

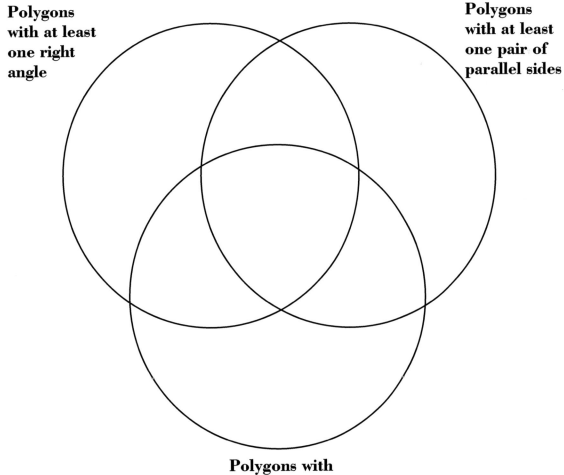

Polygons with at least one right angle

Polygons with at least one pair of parallel sides

Polygons with at least two congruent angles

Name _____   Date _____

Venn Logic 2

At Midwest High School, 31 students participate in the baseball, soccer, and tennis programs. Some play only one of the sports, some participate in two, and a few play all three. Using the clues below, determine how many belong in each category. Place your answers in the most appropriate section of the Venn Diagram.

- 19 students play baseball
- 16 students play soccer
- 17 students play tennis
- 9 students play both baseball and soccer
- 10 students play soccer and tennis
- 8 students play baseball and tennis
- 6 students play all three sports

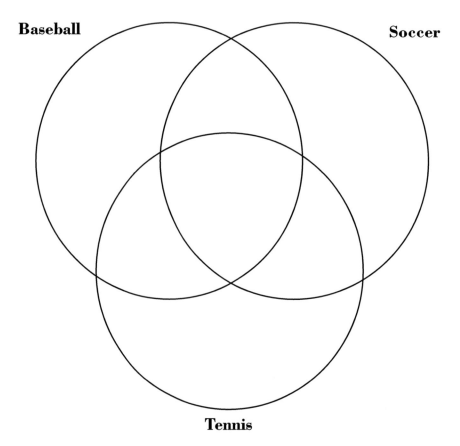

Name_____ Date_____

Venn Logic 3

Directions: Use the diagram below to answer the questions.

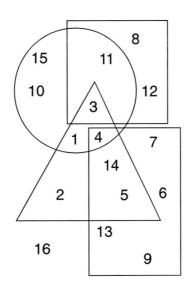

1. What number(s) are in both the triangle and the circle?

2. What number(s) are in the circle, square, triangle, and rectangle?

3. What number(s) are in the circle and square only?

4. What number(s) are in only the circle? only the square?

5. What number(s) are not in the triangle?

6. What number(s) are in none of the polygons?

7. What number(s) are in the triangle or the circle?

8. What number(s) are in the rectangle or the square?

9. Are there any numbers in both the rectangle and the square?

© 1998 J. Weston Walch, Publisher   23   10-Minute Critical-Thinking Activities for Math

PART 3: Number Theory and Problem Solving

The NCTM standards list "Mathematics as Problem Solving" as the first of the standards at all grade levels. Seeing mathematics as more than a collection of rules and algorithms is essential if students are to understand the power and usefulness of math. Students need experiences that (1) give them practice applying varied strategies to solve problems, (2) provide a diversity of problems, and (3) allow them to develop into confident problem solvers. The problems presented in this section offer students a variety of problems and allow for creativity in solution.

The first two "Hit the Target" activities encourage students to practice computation skills while also developing problem-solving strategies. Students may find that making a table will help in the solution of "Hit the Target 3," a strategy useful with other similar problems.

The next three problems are open-ended, and students should be encouraged to find as many solutions to each as they can. In "The 5 by 5 Array," the various solutions are, for the most part, transformations of the original. Students can be introduced to reflections and rotations at this time. "Marbles" is limited only by the number of moves students are allowed to use. Encourage students to share their strategies and solutions with the class. "Triangular Sums" is a problem with many solutions, each one a rotational transformation of the others. Again, encourage students to find as many solutions as they can.

"A Moving Experience" activities are variations of an *oldie but goodie* in the annals of problem solving. Students are given the minimum number of moves necessary to accomplish each task and should be encouraged to meet the challenge.

The NCTM standards recommend on page 95 that the teaching of computation skills (1) foster a solid understanding of and proficiency with simple calculations and (2) give students experiences with problems that will develop estimation techniques. The four "Card Tricks" exercises motivate students to work on open-ended problems that will help strengthen their computation and number-theory skills. As these problems have many different solutions, students should be given the opportunity to share theirs with the class. In this way the myth that there is *one right way* to solve a problem can be dispelled.

Each of the warm-ups in this section can be expanded upon, and additional puzzles can be developed by the teacher or students for use at another time.

(continued)

Name _____ Date _____

Hit the Target 1

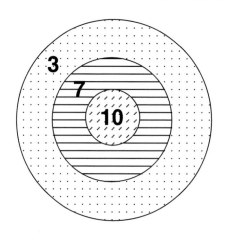

You have six shots at the target above. Show how you can achieve these scores:

⇨ 43 _____

⇨ 40 _____

⇨ 33 _____

⇨ 29 _____

Name _____ Date _____

Hit the Target 2

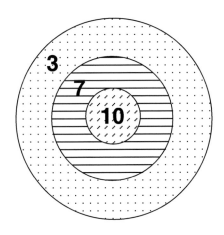

Charlie had six shots at the target above. Which of the following scores could Charlie **not have gotten?**

50   **51**   **52**   **53**

Explain your choice.

_____

_____

_____

_____

_____

_____

_____

© 1998 J. Weston Walch, Publisher   27   10-Minute Critical-Thinking Activities for Math

Hit the Target 3

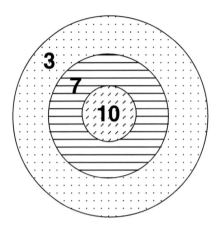

You have six shots at the target shown above. How many different scores are possible if you hit the 10 at least once? _____

How many different scores are possible if you miss the 10 altogether? _____

(*Hint:* Use a table to organize the data.)

The 5 by 5 Array

Can you place the shapes above in the 5 by 5 array so that no shape appears more than once in any row, column, or diagonal?
There may be more than one solution . . . so try to find as many as you can.

Marbles

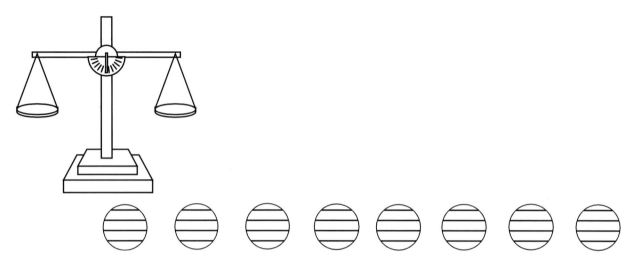

You have eight marbles that are exactly the same size, but one of them weighs slightly more than the others. You cannot tell the difference by just holding them, so you decide to use a balance scale like the one shown above.

Can you find the heavier marble by taking just two weighings? It is possible. See if you can figure out how, and explain your strategies.

Triangular Sums

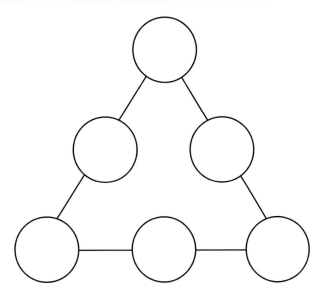

Place the numbers 20, 21, 22, 23, 24, and 25 in the circles so that the sum of each side of the triangle is 67.

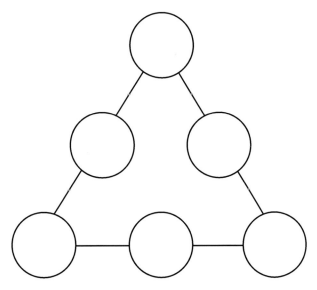

Now rearrange the numbers so that each side adds up to 68.
See how many solutions you can find for each problem.

Name _____ Date _____

A Moving Experience 1

Place two pennies and two nickels on five squares as shown. Your job, should you choose to accept it, is to move the pennies into the places occupied by the nickels and vice versa. But wait—it's not that easy!

The rules follow:

1. The pennies can be moved **only** to the right and the nickels can be moved **only** to the left.

2. You can jump over a coin, but you cannot remove it from the board. And, you can jump only **one** coin at a time or move one square at a time.

3. This can be done in as few as eight moves. How many did it take you?

Move 1: _____

Move 2: _____

Move 3: _____

Move 4: _____

Move 5: _____

Move 6: _____

Move 7: _____

Move 8: _____

Move 9: _____

Move 10: _____

A Moving Experience 2

Place three pennies and three nickels on seven squares as shown. Your job, should you choose to accept it, is to move the pennies into the places that the nickels occupy and vice versa. But wait—it's not that easy!

The rules follow:

1. The pennies can be moved **only** to the right, and the nickels can be moved **only** to the left.
2. You can jump over a coin, but you cannot remove it from the board. And, you can jump only **one** coin at a time or move one square at a time.
3. This can be done in as few as 15 moves. How many did it take you?

Move 1: _____   Move 11: _____

Move 2: _____   Move 12: _____

Move 3: _____   Move 13: _____

Move 4: _____   Move 14: _____

Move 5: _____   Move 15: _____

Move 6: _____   Move 16: _____

Move 7: _____   Move 17: _____

Move 8: _____   Move 18: _____

Move 9: _____   Move 19: _____

Move 10: _____   Move 20: _____

Name _____ Date _____

Card Tricks 1

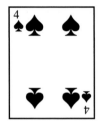  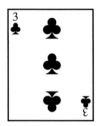 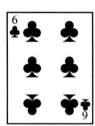

Use the cards above to find answers to these problems. There may be more than one way to solve each problem. Place the numbers in the rectangles and the operation signs in the circles. You may use addition, subtraction, multiplication, and/or division to solve each problem. **But remember the order of operations!** First you must do everything in parentheses, then all multiplication and division (from left to right), and then all addition and subtraction (again from left to right). Use parentheses only when they are needed.

☐ ○ ☐ ○ ☐ ○ ☐ = **4**

☐ ○ ☐ ○ ☐ ○ ☐ = **10**

☐ ○ ☐ ○ ☐ ○ ☐ = **14**

☐ ○ ☐ ○ ☐ ○ ☐ = **24**

Card Tricks 2

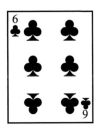

Use the cards above to find answers to these problems. There may be more than one way to solve each problem. Place the numbers in the rectangles and the operation signs in the circles. You may use addition, subtraction, multiplication, and/or division to solve each problem. **But remember the order of operations!** First you must do everything in parentheses, then all multiplication and division (from left to right), and then all addition and subtraction (again from left to right). Use parentheses only when they are needed.

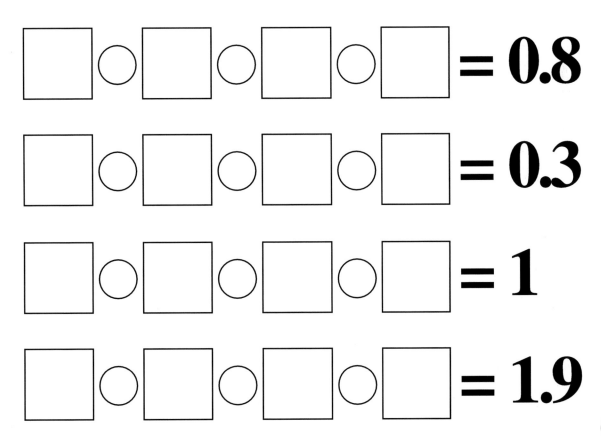

Name _____   Date _____

Card Tricks 3

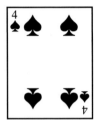   

Use the cards above to find answers to these problems. There may be more than one way to solve each problem. Place the numbers in the rectangles and the operation signs in the circles. You may use addition, subtraction, multiplication, and/or division to solve each problem. **But remember the order of operations!** First you must do everything in parentheses, then all multiplication and division (from left to right), and then all addition and subtraction (again from left to right). Use parentheses only when they are needed. The queen is worth 12.

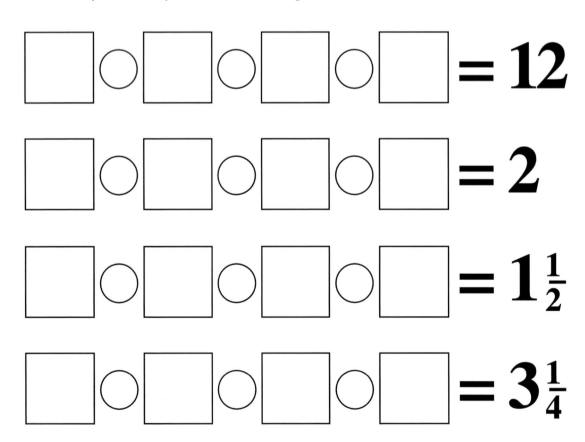

PART 4: Sequences, Patterns, and Codes

The ability to discern patterns enables students to expedite the problem-solving process and helps develop an appreciation for the power of mathematics. Adding a series of numbers can be drudgery; discovering that the sum of the numbers in a 3 × 3 calendar matrix equals the middle number times nine is an exciting revelation, resulting in an *Aha!* experience.

Sequences can be numerical or visual. "Sequences" and "Number Patterns" invite students to find the missing numbers in a sequence of numbers, while "What Comes Next?" develops spatial visualization skills. While answers are provided for these problems, some students may see different solutions. If they can justify their reasoning, these alternate solutions should be encouraged and valued.

Codes and ciphers have fascinated mathematicians for years. Most are based upon number patterns, but some are decoded by discovering the visual patterns built into the code. "Crack the Code 1" is based upon a numerical pattern, the prime numbers. The first prime number is assigned the letter *A*, the second prime number the letter *B*, and so on. After students have solved the riddle, they are asked to figure out how the numbers and letters relate to one another. This is an important part of the problem. The other two "Crack the Code" activities are based upon visual patterns. Students are encouraged to develop their own codes, their own visual patterns, and share them with the class.

Two exercises in this chapter give students the opportunity to investigate patterns in tables: "Calendar Math" and "Addition Table Patterns." Both of these activities ask students to find the sum of the numbers in a square matrix. While doing so, students discover that they don't need to add all of the numbers to find the sum. A pattern emerges that makes problem solving a simpler and more powerful experience.

"Groups of Numbers" activities challenge students to see the relationships among groups of numbers: Are they all multiples of 5 or 7? Are they prime numbers? Students are asked to find what the numbers have in common, to describe the relationship in words, and, finally, to find two more numbers that fit the same pattern.

Name _____ Date _____

Sequences

Can you find a pattern to this sequence? Fill in the missing numbers.

1, 1, 2, 3, 5, ☐, ☐, ☐, 34

Describe the sequence:

Name _____   Date _____

What Comes Next? 1

Examine each of the sequences below. Try to discover the pattern, and then complete the sequence by filling in the empty spaces.

①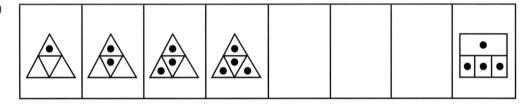

②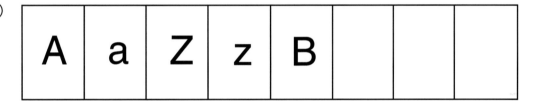

③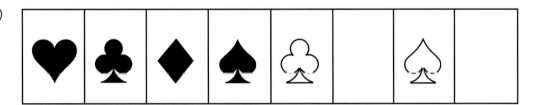

© 1998 J. Weston Walch, Publisher   10-Minute Critical-Thinking Activities for Math

Name _____ Date _____

What Comes Next? 2

Examine each of the sequences below. Try to discover the pattern, and then complete the sequence by filling in the empty spaces.

①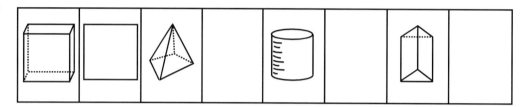

②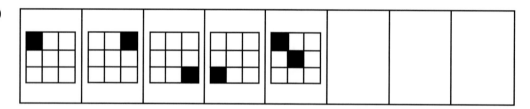

③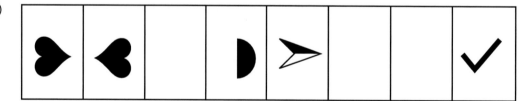

© 1998 J. Weston Walch, Publisher   40   10-Minute Critical-Thinking Activities for Math

Name _____   Date _____

Number Patterns 1

Examine each sequence of numbers carefully. Find the last three numbers to continue the pattern.

① 2, 5, 11, 23, 47, _____, _____, _____

② 2, 2, 4, 6, 10, 16, _____, _____, _____

③ −2, 4, −8, −2, 4, 10, _____, _____, _____

④ 9, 4, 16, 5, 25, 6, 36, _____, _____, _____

Design a sequence of numbers of your own, and share it with the class.

_____

Number Patterns 2

Examine each sequence of numbers carefully. Find the last three numbers to continue the pattern.

① 1, 3, 6, 10, 15, _____, _____, _____

② 2, 10, 5, 25, 20, 100, _____, _____, _____

③ 5, 10, 17, 26, _____, _____, _____

④ 3, –9, –10, 30, 29, –87, –88, _____, _____, _____

Design a sequence of numbers of your own, and share it with the class.

_____

Crack the Code 1

Some of the code we will use is shown below. Can you figure out how we assigned numbers to the letters in the alphabet? Complete this table to find the answer to this riddle:

What didn't Adam and Eve have that everyone else has?

A	B	C	D	E	F	G	H	I	J	K	L	M
2	3	5	7	11	13	17	19	23				
N	O	P	Q	R	S	T	U	V	W	X	Y	Z

__ __ __ __ __ __ __
53 2 61 11 43 71 67

Work with a partner to design your own code.

Crack the Code 2

The code we will use is shown below. Use this table to find the answer to this riddle:

What did one angel say to another?

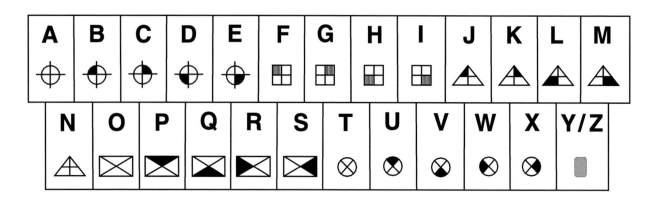

Work with a partner to design your own code.

Crack the Code 3

Inspect the diagrams below. They display a very interesting code. See if you can use the information to find the answer to the riddle below.

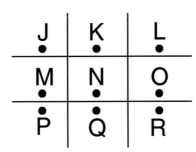

What's more earth-shattering than an elephant playing hopscotch?

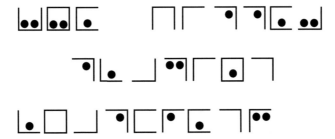

___ ___ ___   ___ ___ ___ ___ ___ ___

___ ___ ___ ___ ___ ___ ___

___ ___ ___ ___ ___ ___ ___ ___

Name _____   Date _____

Calendar Math

Did you know that there are some very interesting patterns in the calendar? Let's look at the one shown below. Choose any nine numbers in a 3 × 3 array, for example, the one outlined. What is the sum of the nine numbers in this array? Add them up: 1 + 2 + 3 + 8 + 9 + 10 + 15 + 16 + 17 = 81. Choose another 3 × 3 grid of numbers; add them up. Continue to do this until you see a pattern emerge. Is there a way to know what the sum is other than adding up all of the numbers?

JANUARY

S	M	T	W	T	F	S
1	2	3	4	5	6	7
8	9	10	11	12	13	14
15	16	17	18	19	20	21
22	23	24	25	26	27	28
29	30	31				

Addition Table Patterns

+	1	2	3	4	5	6	7	8	9
1	2	3	4	5	6	7	8	9	10
2	3	4	5	6	7	8	9	10	11
3	4	5	6	7	8	9	10	11	12
4	5	6	7	8	9	10	11	12	13
5	6	7	8	9	10	11	12	13	14
6	7	8	9	10	11	12	13	14	15
7	8	9	10	11	12	13	14	15	16
8	9	10	11	12	13	14	15	16	17
9	10	11	12	13	14	15	16	17	18

Let's examine some interesting patterns in the addition table. Find the sum of the numbers in the outlined 3 × 3 square: 5 + 6 + 7 + 6 + 7 + 8 + 7 + 8 + 9 = _____.

Pick two more 3 × 3 arrays of numbers. What are their sums? _____
Do you see a pattern? _____

(*Hint:* How many numbers are there in the entire array? Is this number a factor of the sum? Try out this theory using a 4 × 4 array; a 5 × 5 array.)

Name _____ Date _____

Groups of Numbers 1

Each of the shapes below contains a set of numbers that have two attributes in common. (1) Find what the numbers have in common, (2) describe it, and (3) provide two more numbers that share the same things.

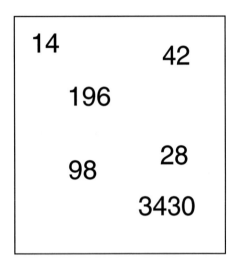

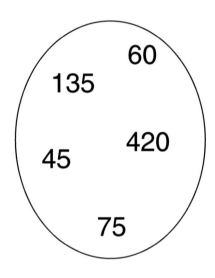

_____   _____
_____   _____
_____   _____

<!-- triangle with 30, 120, 50, 70, 10 -->

_____
_____
_____

© 1998 J. Weston Walch, Publisher   48   10-Minute Critical-Thinking Activities for Math

Name _____ Date _____

Groups of Numbers 2

Each of the shapes below contains a set of numbers that have an attribute in common. (1) Find what the numbers have in common, (2) describe it, and (3) provide two more numbers that share the same thing.

Square: 1, 49, 144, 121, 16, 196

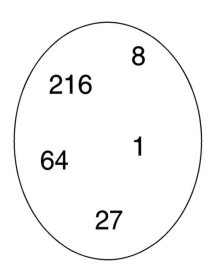

_____
_____
_____

_____
_____
_____

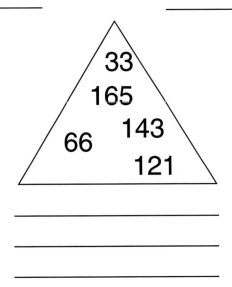

_____
_____
_____

© 1998 J. Weston Walch, Publisher — 10-Minute Critical-Thinking Activities for Math

PART 5: Visual/Geometric Patterns

Many interesting problems show patterns developing from visual representations of data. Looking at problems from a different viewpoint can make what appears to be uninteresting into a fascinating example of a numerical pattern. For example, how many squares are there on a checkerboard? A first response might be "64," since the board has 8 squares per side. But that is only part of the answer: There are 64 squares only 1 square large. What about the squares that are 2×2? And are there overlapping squares? Now we have an interesting problem that encourages students to reason and analyze their options. This is critical thinking disguised as "The Checkerboard."

"The Checkerboard" is but one in a series of four "How many . . ." problems that also includes the following: "How Many Triangles?"; "How Many Triangles in the Pentagon?"; and "How Many Lines in This Hexagon?" Each of these problems helps students develop effective ways to organize visual data and recognize patterns.

"How Many Lines in This Hexagon?" is a visual variation of an old problem sometimes called the Handshake Problem: If each person in a room shakes the hand of every other person in the room, how many handshakes will there be? An organized solution requires the problem solver to visualize counting every handshake of the first person, one less handshake for the next person, and so on. So, if 20 people are in the room, the first person shakes 19 hands, you count 18 of the second person's handshakes (having already counted the handshake between the first and second person), and the problem looks like this:

$$19 + 18 + 17 + 16 + 15 + 14 + 13 + 12 + \ldots 3 + 2 + 1 =$$

In the hexagon five lines radiate from point A, you count four from point B, etc. Ask students if they can develop a strategy to determine the number of diagonals in an octagon, or a decagon, or a dodecagon, and so on. Can a generalization or rule be developed?

"How Many Triangles" gives this theme a slightly different twist. In this problem students must explore regions that contain one triangle, two triangles, and so forth—much like finding the squares in the checkerboard. Thus, students revisit an old problem in a new way.

"How Many Triangles in the Pentagon?" asks students to divide the pentagon into triangular regions (regions of one triangle, two triangles, etc.) and use a table to find the total number in an organized manner. Encourage students to develop other strategies to calculate the number of triangles. Ask them, "Are you sure you have found **all** of the triangles? How can you be sure?" By analyzing their thinking, they are employing the metacognition required to develop critical-thinking skills.

"The Puzzling Cube" activities challenge students to use their critical-thinking skills to visualize what happens in a different kind of problem. Most students have difficulty visualizing cubic nets, or two-dimensional models of cubes. What happens when they are folded up? Will they form a cube? If there are symbols on the faces, how will these symbols appear once a net is folded into a cube? In these activities students are asked to visualize what happens to the net when it becomes a cube. Encourage students to design their own nets, and develop a file of "puzzling cubes" to revisit later.

The next set of problems calls upon similar visualization skills. When we fold a piece of paper and then cut it in some way, it is often a challenge to predict what will happen when we

(continued)

unfold it. Such activities as "Paper Folds" require students to use both spatial visualization and critical thinking skills. While students may at first find this type of problem frustrating, continued exposure and revisiting will improve facility and strengthen visual/spatial intelligence.

Figurate, or geometric, numbers are an interesting juxtaposition of number and shape. Square numbers (1, 4, 9, 16, etc.) can be used to form squares, and numbers that can be used to form triangles (1, 3, 6, 10, etc.) are called triangular numbers. "Patterns in Geometric Numbers" introduce the concept of figurate numbers and then look for patterns that develop from their first and second differences. Students should be encouraged to analyze and explain the patterns they find.

The tangram, a seven-piece puzzle, consists of one square, one parallelogram, and two small, one medium, and two large isosceles right triangles. Since the medium triangle, square, and parallelogram have the same area, these shapes can be used for interesting discussions on area. The triangles can be used to discuss the concepts of similarity and congruence. This is a wonderful puzzle to help students analyze shapes and discover relationships. "Tangram Puzzler" activities will help students become more familiar with geometric terms, develop visual/spatial skills, and hone their critical-thinking skills. A tangram puzzle is supplied below to solve the puzzles and to make up new ones to challenge the class. Students may be encouraged to cut up the puzzles on the reproducible pages.

Tangram Puzzler

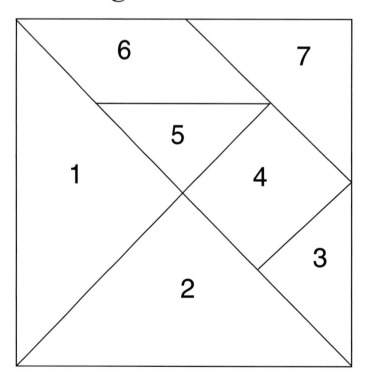

Carefully cut out seven tangram pieces to help you solve the "Tangram Puzzler" activities. You can also use them to create your own puzzlers to challenge the class.

How Many Triangles?

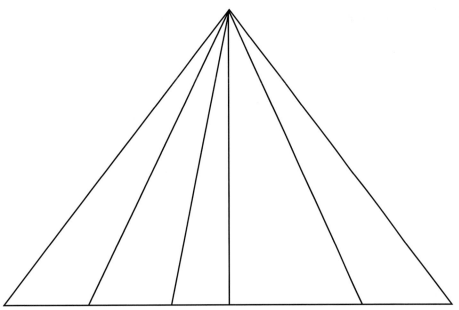

How many triangles can you find in this figure? You may find organizing your thoughts easier if you label each small triangle and use a table.

The Checkerboard

How many squares can you find on a checkerboard? There are a great many more than 64! Work to find the total number of squares. Perhaps using a table would help. Then, describe the pattern you found.

How Many Triangles in the Pentagon?

How many triangles can you find in this pentagon? Dividing it into regions, numbering each triangle, and using a table may make it easier for you to organize your data.

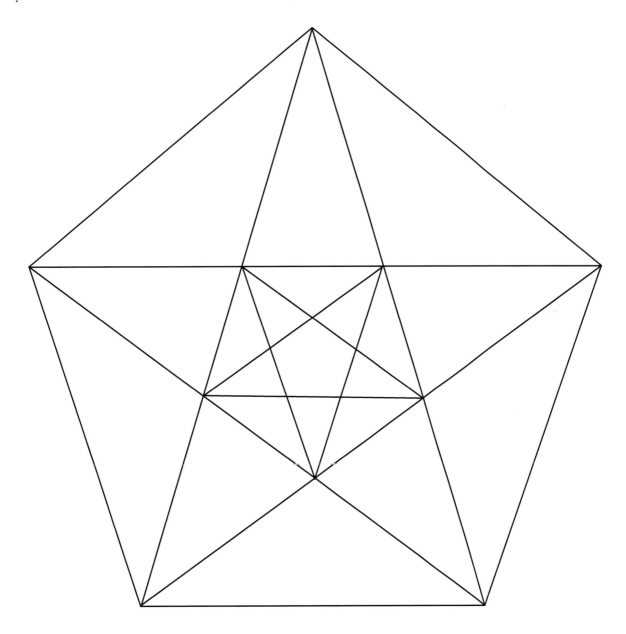

How Many Lines in This Hexagon?

How many lines are there between the points on the edge of the circle? Do you think there might be a pattern?
(*Hint:* Try to discover a pattern by counting the lines from A, then B, etc.)

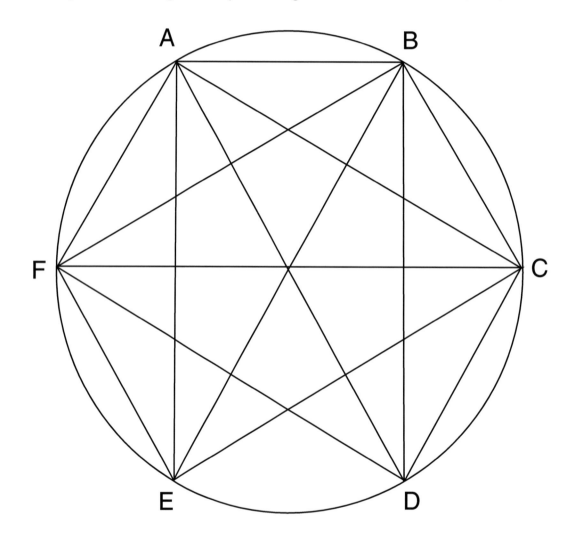

Name _____   Date _____

The Puzzling Cube 1

A cubic net looks like this: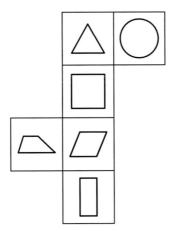

What does it look like after it has been folded into a cube: A, B, C, or D?

A.

B.

C.

D.

© 1998 J. Weston Walch, Publisher   57   10-Minute Critical-Thinking Activities for Math

Name_____ Date_____

The Puzzling Cube 2

A cubic net looks like this: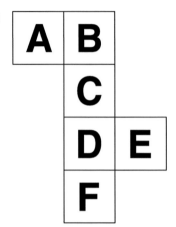

What does it look like after it has been folded into a cube: A, B, C, or D?

A.

B.

C.

D.

Paper Folds 1

Take a square of paper:

Fold it in half:

Fold it in half again and punch out a design:

What will it look like when it is unfolded?

Directions: The papers below have been folded like the one above. Your job is to figure out what they will look like when they are unfolded—to move from step 3 to step 4!

Folded paper **My drawing of it unfolded**

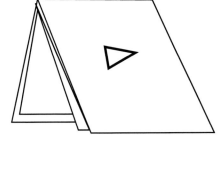

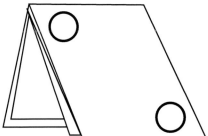

Design your own paper fold puzzle, and share it with the class.

Name _____   Date _____

Paper Folds 2

Take a square of paper:

Fold it in half:

Fold it in half again and punch out a design:

What will it look like when it is unfolded?

Directions: The papers below have been folded like the one above. Your job is to figure out what they will look like when they are unfolded—to move from step 3 to step 4!

Folded paper   **My drawing of it unfolded**

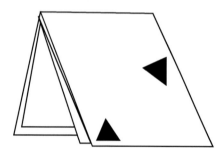

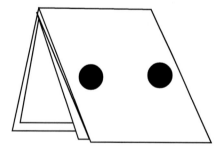

Design your own paper fold puzzle, and share it with the class.

Patterns in Geometric Numbers 1

The numbers shown below are called triangular numbers. Four of them have been sketched for you. Copy the triangles shown. Then do the following:

- Sketch the next one in the series.
- Count the number of dots in each triangular figure.
- The differences in their numbers have been figured for those shown.
- The second difference has been shown for the first two figures; figure the second differences for the next two.

Can you use these number patterns to discover the number of dots in the fifth? . . . the tenth? . . . the *n*th?

Explain the pattern you discovered.

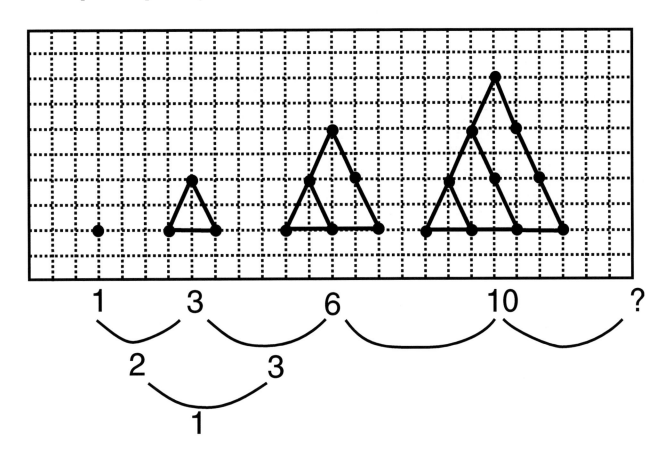

Patterns in Geometric Numbers 2

The numbers shown below are called square numbers. Four of them have been sketched for you. Copy the squares shown. Then do the following:

- Sketch the next one in the series.
- Count the number of dots in each square figure.
- The differences in their numbers have been figured for those shown.
- The second difference has been shown for the first two figures; figure the second differences for the next two.

Can you use these number patterns to discover the number of dots in the fifth? . . . the tenth? . . . the nth?

Explain the pattern you discovered.

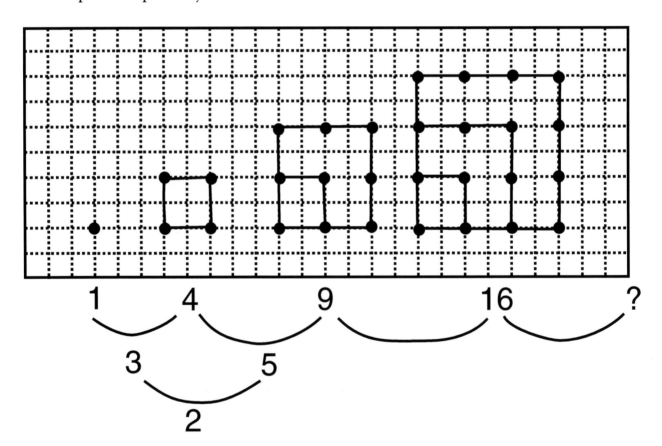

Tangram Puzzler 1

Use three triangles from the tangram puzzle to form each of the following:

1. larger triangle
2. rectangle that's not a square
3. parallelogram
4. square

Sketch a picture of each solution. Use the number assigned to each piece in the tangram below to explain your drawing. See how many different shapes you can find.

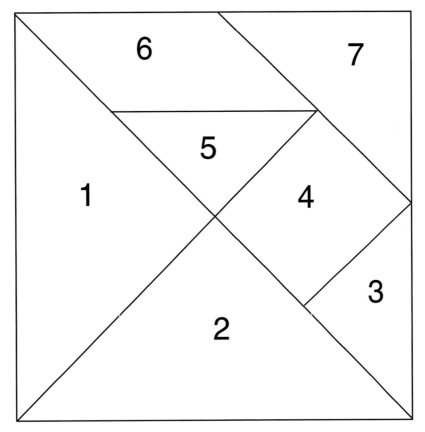

The Tangram

Name _____   Date _____

Tangram Puzzler 2

Combine the two quadrilaterals from the tangram puzzle with one of the small triangles to form two different trapezoids with the same area.

1. How can you prove these two trapezoids have the same area?
2. Are they congruent? Explain your answer.

Sketch a picture of each solution. Use the number assigned to each piece in the tangram below to explain your drawing. See how many different trapezoids you can find.

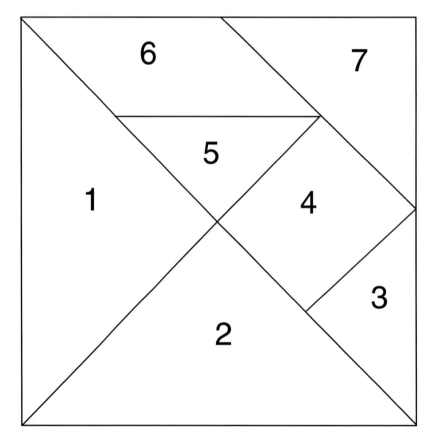

The Tangram

© 1998 J. Weston Walch, Publisher   64   *10-Minute Critical-Thinking Activities for Math*

Answers

He Said—She Said (pp. 2-3)
1. Possible rule is $2n + 1$
2. Possible rule is $2n - 4$

What Comes Next? 1 (p. 4)
1. Twentieth = 41 toothpicks
2. Fiftieth = 101 toothpicks
3. One-hundredth = 201 toothpicks
4. Possible answer: $2n + 1$

What Comes Next? 2 (p. 5)
1. Twentieth = 41 squares
2. Fiftieth = 101 squares
3. One-hundredth = 201 squares
4. Possible answer: $2n + 1$

What Comes Next? 3 (p. 6)
1. Twentieth = 62 squares
2. Fiftieth = 152 squares
3. One-hundredth = 302 squares
4. Possible answer: $3n + 2$

What Comes Next? 4 (p. 7)
1. Twentieth = 84 squares
2. Fiftieth = 204 squares
3. One-hundredth = 404 squares
4. Possible answer: $4n + 4$

The Ladybug Concert (p. 8)

A possible algebraic solution would look like this:

Let n = the shortest line
$n + (n + 1) + (n + 2) + (n + 3) + (n + 4) + (n + 5) = 327$
$6n + 15 = 327$ so $6n = 312$ and $n = 52$

Therefore, the lines contain 52, 53, 54, 55, 56, and 57 people. It is possible for students to use entirely different strategies, like finding the average number of people in each line (54.5) and using trial and error to find the answer. Be sure students have an opportunity to share their thinking and problem-solving strategies.

The Peanut Problem (p. 9)

One possible solution using algebra:

Let c = number of cashews; $2c$ = number of Brazil nuts and almonds; and $4c$ = number of peanuts.
$2c + 4c + c + 2c = 81$

Using transformations: $c = 9$; 9 cashews, 18 each of Brazil nuts and almonds, and 36 peanuts.

Number Magic 1 (p. 10)

Steps are:

1. n
2. $n + 2$
3. $4(n + 2) = 4n + 8$
4. $\dfrac{4n + 8}{2} = 2n + 4$
5. $2n + 4 - 4 = 2n$
6. $\dfrac{2n}{n} = 2$

Number Magic 2 (p. 11)

Steps are:

1. n
2. $6n$
3. $6n + 4$
4. $\dfrac{6n + 4}{2} = 3n + 2$
5. $3n + 2 - 2 = 3n$
6. $\dfrac{3n}{3} = n$

Number Magic 3 (p. 12)

Steps are:

1. n
2. $n + 5$
3. $5(n + 5) = 5n + 25$
4. $5n + 25 - 25 = 5n$
5. $\dfrac{5n}{n} = 5$

Ladder Logic (p. 14)

1. Possible solution: point, paint, pains, rains, raids
2. Possible solution: bolt, colt, cold, cord, card

Dinner Table Logic (p. 15)

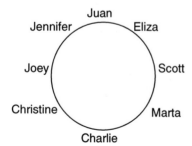

Fun with Logic 1 (p. 16)

	Architect	Botanist	Chemist	Doctor	Electrician
Amy	X	X	X	X	O
Burt	O	X	X	X	X
Charlie	X	O	X	X	X
Donna	X	X	O	X	X
Eliza	X	X	X	O	X

Fun with Logic 2 (p. 17)

	Art	Biology	Math	Social Studies	Physical Education
Anthony	X	O	X	X	X
Brandy	O	X	X	X	X
Molly	X	X	X	X	O
Sam	X	X	O	X	X
Polly	X	X	X	O	X

Fun with Logic 3 (p. 18)

	Monday	Tuesday	Wednesday	Thursday	Friday
Pat	X	X	O	X	X
Charlie	X	O	X	X	X
Luan	O	X	X	X	X
Maria	X	X	X	O	X
Josie	X	X	X	X	O

Splatland 1 (p. 19)

Dib is 4; Dab is 8; and Dud is 2

Splatland 2 (p. 20)

Dib is 9; Dud is 3; and Doozie is 18

Venn Logic 1 (p. 21)

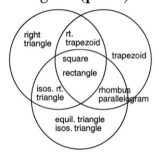

Venn Logic 2 (p. 22)

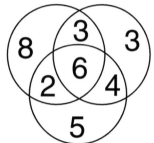

Venn Logic 3 (p. 23)

1. 1, 3, 4
2. none
3. 11
4. 10 and 15; 8 and 12
5. 10, 15, 11, 8, 12, 7, 6, 13, 9, 16
6. 10, 15, 16
7. 1, 2, 3, 4, 5, 10, 11, 14, 15
8. 3, 8, 11, 12, 4, 5, 6, 7, 9, 13, 14
9. no

Hit the Target 1 (p. 26)

Possible answers:

43 = three 10's, two 7's, and one 3
40 = two each of 10's, 7's, and 3's
33 = one 10, two 7's, and three 3's
29 = one 10, one 7, and four 3's

Hit the Target 2 (p. 27)

Charlie could not have gotten a score of 52; all the others are possible.

50 = four 10's, one 7, and one 3
51 = three 10's and three 7's
53 = five 10's and one 3

Hit the Target 3 (p. 28)

There are 21 possible scores including at least one 10: six scores with one 10, five with two 10's, four with three 10's, etc., for 6 + 5 + 4 + 3 + 2 + 1 = 21 possible scores.

Seven scores are possible if you miss the 10 altogether.

The 5 by 5 Array (p. 29)

See student work.

Marbles (p. 30)

Place three marbles on each side of the scale. If the scale remains balanced, then weigh the remaining two to determine which one is heavier. If the scale tips to one side, pick any two of the marbles from that side and place them on the scale. If the scale tips to one side, you know which marble is heavier. If the scale remains balanced, then the marble that is left is the heavier one.

Triangular Sums (p. 31)

Possible solutions:

Sum of 67

```
        (20)
     (25)  (23)
  (22)  (21)  (24)
```

Sum of 68

```
        (21)
     (24)  (22)
  (23)  (20)  (25)
```

A Moving Experience 1 & 2 (pp. 32–33)

The minimum number of moves required is described in each activity.

Card Tricks 1, 2, & 3 (pp. 34–36)

These are only possible solutions among the **many** ways to solve these problems.

Card Tricks 1
$(8 \div 4) + (6 \div 3) = 4$
$(8 - 3) \times (6 - 4) = 10$
$(8 + 4) + (6 \div 3) = 14$
$(6 \div 3) \times (8 + 4) = 24$

Card Tricks 2
$(8 + 6 - 10) \div 5 = 0.8$
$(6 + 5 - 8) \div 10 = 0.3$
$(10 \div 5 + 6) \div 8 = 1$
$(6 + 5 + 8) \div 10 = 1.9$

Card Tricks 3
$[6 \div (12 \div 4)] + 10 = 12$
$(12 - 10 + 6) \div 4 = 2$
$(4 \div 6) + (10 \div 12) = 1\frac{1}{2}$
$(12 \div 16) + (10 \div 4) = 3\frac{1}{4}$

Sequences (p. 38)

8, 13, 21—This is Fibonacci's sequence.

What Comes Next? 1 & 2 (pp. 39–40)

Number Patterns 1 (p. 41)

1. 95, 191, 383 (double the previous number and add 1)
2. 26, 42, 68 (add the two previous numbers)
3. –20, –14, 28 (add 6; multiply by –2)
4. 36, 7, 49 ($\sqrt{n}+1; n^2; \sqrt{n}+1; n^2$; etc., where n is the preceding term)

Number Patterns 2 (p. 42)

1. 21, 28, 36 ($n+6$; $n+7$; $n+8$, where n is the preceding term; or $(n^2+n)/2$, where n is the number of the term)
2. 95, 475, 470 (multiply by 5; subtract 5 from product)
3. 37, 50, 65 (6^2+1; 7^2+1; 8^2+1)
4. 264, 263, –789 (multiply by –3; subtract 1 from product)

Crack the Code 1 (p. 43)

These are all prime numbers. Riddle: PARENTS

Crack the Code 2 (p. 44)

Riddle: HALO

Crack the Code 3 (p. 45)

Riddle: TWO HIPPOS PLAYING LEAPFROG!

Calendar Math (p. 46)

The pattern in a 3 × 3 array: 9 × the center number = the sum of all the numbers in that array.

Addition Table Patterns (p. 47)

The pattern in a 3 × 3 array: 9 × the center number = the sum of all the numbers in that array. In larger $n \times n$ arrays, the sum of all the numbers = the number of squares in the array × the number that repeats diagonally from top right to bottom left (the center number in $n \times n$ arrays where n = an odd number).

Groups of Numbers 1 (p. 48)

Square = multiples of 7 and 2

Oval = multiples of 5 and 3

Triangle = multiples of 5, 2, and 10

Groups of Numbers 2 (p. 49)

Square = square numbers

Oval = cubic numbers

Triangle = multiples of 11

How Many Triangles? (p. 53)

The original triangle has been divided into smaller triangles that can be labeled as follows:

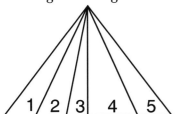

1-part triangles	2-part triangles	3-part triangles	4-part triangles	5-part triangles
1	1-2	1-2-3	1-2-3-4	1-2-3-4-5
2	2-3	2-3-4	2-3-4-5	
3	3-4	3-4-5		
4	4-5			
5				
Totals: 5 +	4 +	3 +	2 +	1 = 15

The Checkerboard (p. 54)

The table students can use to help them organize their data would look like the one below.

The answers are shown in the second column. The pattern is that the number of squares is always a square number, starting with 8^2 and continuing with 7^2, 6^2, etc.

Size of Square	Number of Squares
1 × 1	64
2 × 2	49
3 × 3	36
4 × 4	25
5 × 5	16
6 × 6	9
7 × 7	4
8 × 8	1
TOTAL	204

How Many Triangles in the Pentagon? (p. 55)

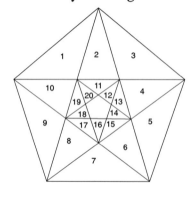

1-part triangles	2-part triangles	3-part triangles
1 6	1-2 2-3	1-2-3
2 7	3-4 4-5	3-4-5
3 8	5-6 6-7	5-6-7
4 9	7-8 8-9	7-8-9
5 10	9-10 10-1	9-10-1
Total: 10 +	10 +	5 = 25

So, 25 is the number of triangles in the outer region of the large pentagon. As the small inner pentagon is divided up the same way, the total number of triangles in both outer and inner regions of the pentagon is 50.

How Many Lines in This Hexagon? (p. 56)

Explore the relationship between this problem and others like it:

5 + 4 + 3 + 2 + 1 = 15

The Puzzling Cube 1 & 2 (pp. 57–58)

1. A   2. C

Paper Folds 1 (p. 59)

Paper Folds 2 (p. 60)

 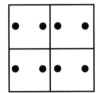

Patterns in Geometric Numbers 1 (p. 61)

The algebraic form used to describe triangular numbers is $\dfrac{n(n+1)}{2}$

Patterns in Geometric Numbers 2 (p. 62)

The algebraic form used to describe square numbers is n^2

Tangram Puzzler 1 & 2 (pp. 63–64)

See student answers.

Share Your Bright Ideas with Us!

We want to hear from you! Your valuable comments and suggestions will help us meet your current and future classroom needs.

Your name_____Date_____

School name_____

School address_____

City_____State_____Zip_____Phone number (_____)_____

Grade level taught_____Subject area(s) taught_____Average class size_____

Where did you purchase this publication?_____

Was your salesperson knowledgeable about this product?   Yes_____   No_____

What monies were used to purchase this product?

    ___School supplemental budget   ___Federal/state funding   ___Personal

Please "grade" this Walch publication according to the following criteria:

Quality of service you received when purchasing .... A   B   C   D   F
Ease of use.. A   B   C   D   F
Quality of content... A   B   C   D   F
Page layout .... A   B   C   D   F
Organization of material .... A   B   C   D   F
Suitability for grade level .... A   B   C   D   F
Instructional value.. A   B   C   D   F

COMMENTS:_____

_____

What specific supplemental materials would help you meet your current—or future—instructional needs?

_____

Have you used other Walch publications? If so, which ones?_____

May we use your comments in upcoming communications?   ___Yes   ___No

Please **FAX** this completed form to **207-772-3105**, or mail it to:

    Product Development, J. Weston Walch, Publisher, P. O. Box 658, Portland, ME 04104-0658

We will send you a **FREE GIFT** as our way of thanking you for your feedback.   **THANK YOU!**